THE OPEN U
MATHEMAT
AN INTERFA
MA290: TOPI

BLOCK 1   MATHEMAT

CW01457090

UNIT 4

THE GREEK STUDY OF CURVES

PREPARED BY JOHN FAUVEL FOR THE COURSE TEAM

THE OPEN UNIVERSITY

CONTENTS

This unit forms part of an Open University course. The set book for the course, to which reference is made as **SB**, is:

John Fauvel and Jeremy Gray (editors), *History of Mathematics: A Reader*, Macmillan 1987.

Acknowledgements

Grateful acknowledgement is made to the following sources for material used in this unit: *Figure 1*: N. Lewis, *Papyrology*, Yale Classic Studies XXVIII, © Department of Classics (Yale University, 1985); *Figure 12*: J. Godwin, *Athanasius Kircher* (Thames and Hudson, 1979).

The Open University, Walton Hall, Milton Keynes.

First published 1987. Reprinted 1989

Designed by the Graphic Design Group of the Open University.

Typeset in Great Britain by Santype International Ltd, Salisbury.

Printed in Great Britain by BPCC Wheatons Ltd, Exeter

ISBN 0 335 14248 6

This text forms part of the correspondence element of an Open University Second Level Course.

For general availability of supporting material referred to in this text, please write to Open University Educational Enterprises Limited, 12 Cofferidge Close, Stony Stratford, Milton Keynes MK11 1BY, Great Britain.

Further information on Open University courses may be obtained from The Admissions Office, The Open University, P.O. Box 48, Milton Keynes MK7 6AB.

4.1 THE PROBLEMS OF CURVES

In the previous two units you have seen several examples of curves which were studied, in greater or lesser depth, by Greek mathematicians. The curves varied quite widely, from the basic *circle* to the ingenious *trisectrix*. In this unit we look more coherently at the definitions and properties of curves in Greek times, leading up to the works of the two great—but very different—third century BC geometers Archimedes and Apollonius. This study is both interesting in its own right, and important for us later on because of the critical role these Greek achievements subsequently played. From the seventeenth century onwards, the Greek conception of properties of curves, methods of investigating them, and even the way curves were thought of, were variously built upon or else consciously rejected; but always with a deep debt to the Greek past, and to Archimedes and Apollonius in particular.

What *is* a curve? Or rather, what were curves taken to be in classical Greek times? How were they defined? This is clearly an important question because, as you will have realised after your study of Greek proof in *Unit 3*, the way something is defined is bound to influence, if not determine, what can be proved about it. One interesting aspect of the various curves you have already seen is that their definitions were of quite different kinds.

Question 1 Read Euclid's definition of a *circle* (*Elements* I, Def. 15), and compare it with the definitions of the *parabola* possibly due to Menaechmus (*Unit 3*, Box 4), and that of the *trisectrix* (or *quadratrix*) of Hippias (*Unit 3*, Section 3). Can you see any differences in the *kinds* of definition given?

Comment ————————————————

The trisectrix definition stands out by invoking *motion.* This concept does not arise with the other two. The circle is defined in terms of the property that all straight lines from the centre to the curve have the same length, and the parabola was defined as a plane section of a cone.

There is another, more subtle, way in which the definitions differ. Those of the trisectrix and parabola tell you what to do to produce the curve, in terms (respectively) of intersecting two moving lines and of intersecting a cone and a plane. But the circle is defined in terms of a property that it has, rather than what you do to get one. ■

We could describe these two ways of defining a curve as *definition by genesis* and *definition by property*. Although most curves seem to have been given by genesis, by saying how they were constructed, there are hints that definition by property was considered the better approach to aim for (by Plato, at least). Recall that in the discussion of geometry in *Republic* Socrates said that geometry is

> A science quite the reverse of what is implied by the terms its practitioners use. . . . They talk about 'squaring' and 'applying' and 'adding' and so on, as if they were *doing* something and their reasoning had a practical end, and the subject were not, in fact, pursued for the sake of knowledge. . . . it must, I think, be admitted that the objects of that knowledge are eternal and not liable to change and decay.

But what mathematicians were told they ought to do, and what they actually did, were not necessarily the same. There is not even consistency throughout Euclid's *Elements* on the matter. For instance, his definition of *circle* is given by property, but the definition of *sphere* is by genesis. Have a look at that now (*Elements* XI, Def. 14 in **SB** 3.E2) and then try this question to check that you have understood the distinction between the two sorts of definition.

Question 2 Compare Euclid's definitions of a circle and a sphere, and try to recast each in the style of the other: that is, give a definition of the circle by genesis, and a definition of the sphere by property, using the other as model.

Comment ————————————————

The circle's definition comes out something like this: a circle is the figure traced by the end of a straight line segment moving in the plane so that its other end is always

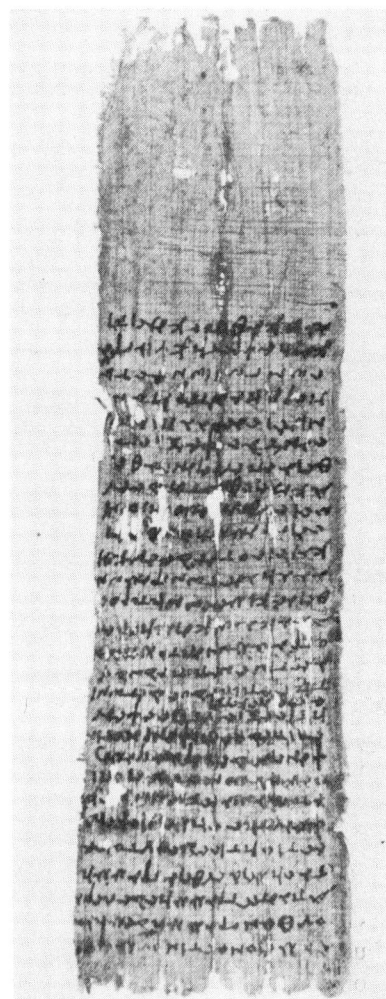

Figure 1 *Elements* I, Defs 1–10. A copy on papyrus made in the third century AD, perhaps by a schoolmaster

This distinction is not found in these explicit terms in the classical sources, but is made for the purpose of historical analysis. Here we are following the discussion of A. G. Molland, 'Shifting the foundations: Descartes' transformation of ancient geometry', *Historia Mathematica*, **3**, (1976), 21–49.

SB 3.E2

The sphere, of course, is a curved surface rather than a curved line, but that difference is not significant here.

fixed. And that for the sphere is something like: a sphere is the surface such that all straight lines falling on it from one particular point (its '*centre*') are equal to one another. (The precise wording does not matter, so long as you picked up the idea that for the sphere some characterising property was sought, and for the circle something you could do to get one.) ■

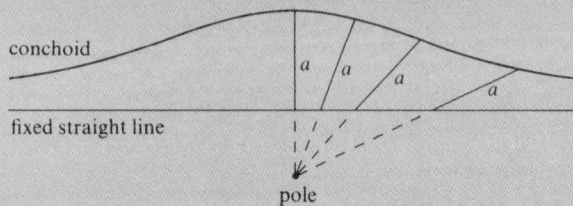

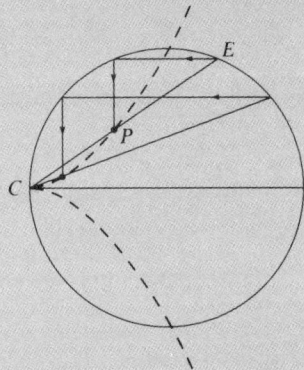

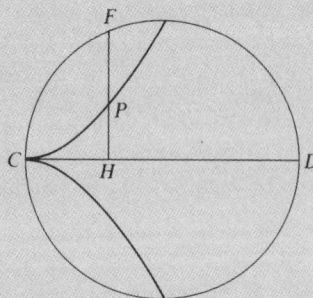

So in these cases, at least, definitions of either kind are equally easy to formulate, although for other curves an appropriate defining property might be harder to enunciate. Many Greek curves came into existence only through being constructed for solving problems—the 'three classical problems' were especially fertile in this respect. Such curves had no other known properties until further investigated. We shall shortly see what kinds of properties of curves were studied, but first we survey what curves were known.

Recall Pappus's classification of problems according to the curves needed in their solutions. This in effect put curves into three classes: *plane* (the circle and the straight line); *solid* (conic sections—parabola, hyperbola and ellipse—formed by plane sections of a cone); and *linear* (anything else, such as the quadratrix, spiral, conchoid and cissoid—see Box 1). This three part classification was to be important in the seventeenth century, when Descartes developed it further. Other curves were known too, beyond those mentioned by Pappus, as we learn from a different classification due to Geminus (c. first century BC) and described by Proclus; see Box 2.

Box 2   Note on Geminus' classification of lines

We learn of this from Proclus, whose text is rather complicated, so what is given here is the diagrammatic summary due to the historian Sir Thomas Heath. The curves mentioned are those given by Proclus.

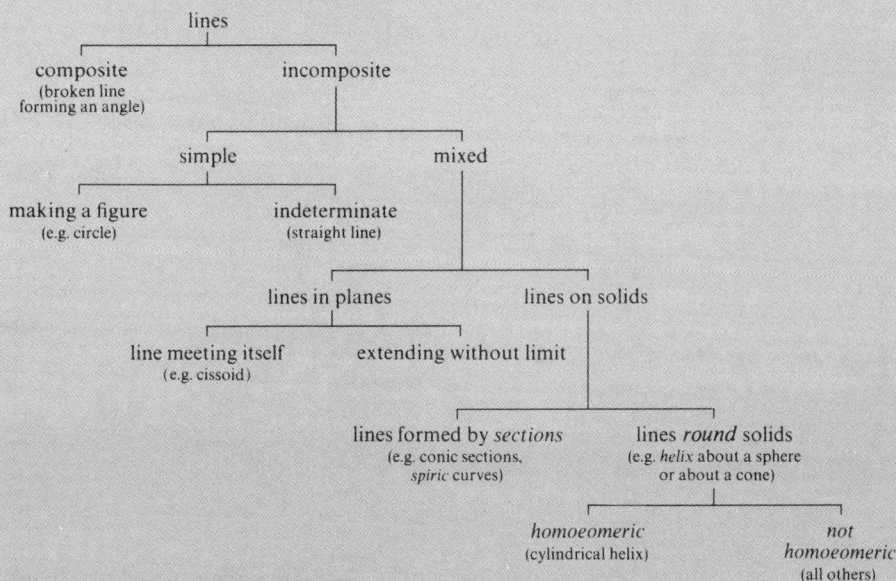

A homoeomeric line is one whose every part is alike, so that any part can be got to coincide with any other; there are three such lines, the circle, straight line and cylindrical helix (the shape of an endless spring). All the other terms are fairly self-explanatory.

Not all of these curves were studied equally intensively, but the way several come down to us linked to the names of individuals suggests that the activity of devising and investigating new curves was the way a mathematician could gain prestige. A case in point might be Perseus (perhaps third century BC), known to us only from references by Proclus many centuries later, who seems to have investigated the *spiric sections*. These are curves, somewhat analogous to conic sections, arising from the intersection of a plane and a *spire* (now called a *torus* or ring doughnut). According to Proclus, Perseus showed for each spiric section 'what its property was'; that is, in terms of our earlier distinction, he produced a definition by property for each section. This is the way Apollonius approached conic sections, as we shall see later.

We now move on from questions of definition and classification, to see what *properties* of curves were studied; that is, considering a curve not as a solution to some other problem, but as an object of geometrical interest in its own right, about which theorems could be proved. Broadly, the properties investigated were analogous to those of figures made up of straight lines: angles and other aspects of lines meeting, line lengths, areas enclosed by curved lines. But all these were more problematical for curves than the equivalent rectilinear (straight-line) investigations. Let us look briefly at the question of *angles* to illustrate this.

Both Hippocrates in the fifth century BC, and Aristotle in the fourth, used the concept of the angle between a straight and a curved line. And Euclid explicitly defined rectilineal angle (*Elements* I, Def. 9) as a sub-class of plane angles (Def. 8), thus implying that the latter could be angles between curved lines. Then in *Elements* III, 16 Euclid included a result concerning these two sorts of angle (Figure 2).

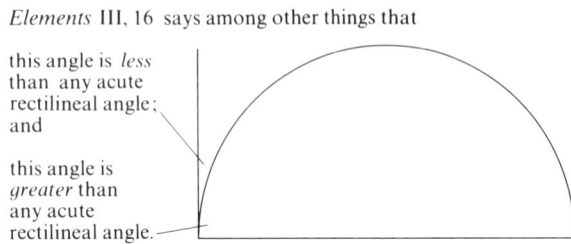

Elements III, 16 says among other things that

this angle is *less* than any acute rectilineal angle; and

this angle is *greater* than any acute rectilineal angle.

Figure 2

Much later, Proclus recorded a classification of angles due to Geminus, every bit as elaborate and complicated as his classification of lines. In the last few centuries BC there is evidence of much discussion of different definitions of angles, and what kind of a mathematical object or property they are. And yet, all this thought and activity notwithstanding, nothing very satisfactory emerged. The difficulties and paradoxes arising out of trying to consider angles between curved lines were perhaps too great. At all events, the critically important question of how to compare the size of two curvilinear angles seems to have made little progress.

See for example Proclus' commentary on *Elements* I, Def. 8, **SB** 3.B1(b).

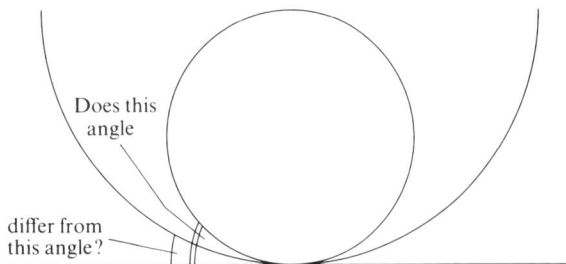

Does this angle

differ from this angle?

Figure 3

Another question concerning the meeting of lines was developed more productively: that of where and how curves *touched*. A straight line which touches a curve is called a *tangent*. *Elements* III, 16 amounts to a construction of the tangent to a circle at any chosen point. Euclid showed that constructing a straight line at right angles to the diameter gives rise to a line with the tangent property, which he defined as meeting the circle but not cutting it (*Elements* III, Def. 3). (Note that without some such reasoning it would not be clear that a tangent existed, that there *was* any line with the defined property.) There are two other properties of tangents to circles which emerge from this proposition. Any particular tangent is unique (not more than one line can be tangent to the circle just there), and the tangent touches the curve at a single point. The latter claim had apparently been the subject of previous controversy, like so much else in Greek geometry. Aristotle, referring to the difference between geometrical lines and visible reality, remarked

> a circle in fact touches a straight-edge, not at one point only, but in the way that Protagoras, in his refutation of the geometers, said that it did.

Aristotle, *Metaphysics* 998ᵃ, in T. L. Heath, *Mathematics in Aristotle* (Oxford, 1949) p. 204.

So in the later fifth century BC (Protagoras was a Sophist teacher contemporary with Socrates) there seem to have been discussions over whether a tangent touches a curve at one point, or at many as sense-evidence suggests.

Problems concerning the length of curved lines, and areas enclosed by them, and volumes enclosed by curved surfaces, were difficult. But in the hands of Archimedes, especially, they were solved with astonishing virtuosity. You saw that such concerns were at the forefront of the geometric research tradition from at least the time of Hippocrates, in the fifth century BC. In the next two sections we look at developments in the finding of areas and volumes from then up to Archimedes, two centuries or so later. To conclude this section, several themes can be seen to come together in an interesting curve we have not met before, the spiral of Archimedes.

Question 3 Browse through the extracts from Archimedes' *On Spirals* (**SB** 4.A7) to form an impression of its content and style. Then consider the following questions:

(i) What is Archimedes' definition of a spiral? Is this a definition by property or by genesis?

(ii) What kind of properties of the curve is he interested in?

Comment

(i) The definition ('Definition 1') involves two simultaneous motions, a straight line revolving while a point moves out along the line. (Both motions are uniform—that is, each is at constant speed.) The spiral is the path of the point in relation to the plane in which the motions are taking place. It is, if you like, the path, relative to the table, of a fly walking straight outwards on a revolving gramophone record.

This is a definition by genesis, because it tells you what to do (in the imagination) to produce the curve. Any properties the curve may have must be deduced from this description of the double motion.

(ii) Archimedes describes the major results in the penultimate paragraph of his introductory letter to Dositheus. Of the four propositions mentioned here, three are to do with areas and one to do with line length in relation to the tangent to the curve. So he is interested in essentially those curve properties we mentioned earlier—tangents, line length, area. ■

As Archimedes explained in the final paragraph of the letter, the other propositions in the book are mostly those needed to establish his major results. In part, their function is to mediate between the genetic definition of his curve and the eventual proofs of the significant properties, for it is interesting that in the latter proofs no concept of motion or time appears. We shall not study the structure of an Archimedean proof in detail until later in the unit, but his results on spirals are so remarkably beautiful and simple that it is worth understanding what they are.

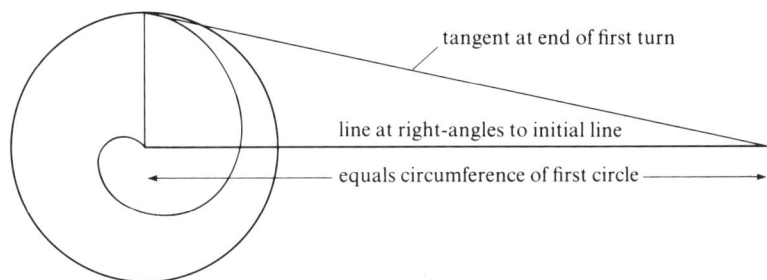

Figure 4

Proposition 18 shows that where the tangent at the end of the first turn cuts the line at right angles to the initial line, the straight-line distance to the origin is equal to the circumference of the first circle (Figure 4). This is a most striking result. A curved line (the circle) can be shown to have the same length as a straight line. Proposition 20 gives a similar result for the tangent at any point on the spiral. Proposition 24 is the important result on areas, that the 'first area' is one-third of the 'first circle' (that is, the shaded area in Figure 5 is one-third of the area within the dotted circle). Again, a simple, elegant, quite unexpected result, of which Archimedes proved analogues for further turns of the spiral.

You should now have some feeling for the sort of enterprise that constituted the Greek study of curves. We shall need to come back to how results were *proved*, which was one of the most impressive, or daunting, legacies of Greek mathematics to sixteenth and seventeenth century Europe. In the next section we look at problems of determining the area bounded by curves, with particular reference to the archetypal problem of squaring the circle.

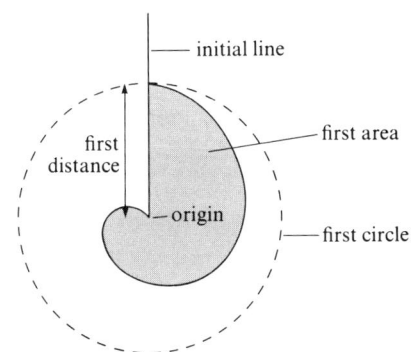

Figure 5

4.2 SQUARING THE CIRCLE

According to Proclus, the problem of squaring the circle arose by analogy, from the early Greek success in determining exactly the area bounded by any figure with straight sides. In his commentary on the major result of the first book of Euclid's *Elements* (I, 45) Proclus observed

Proposition 45 shows that any rectilinear figure can be transformed into a parallelogram of any desired angle.

> It is my opinion that this problem is what led the ancients to attempt the squaring of the circle. For if a parallelogram can be found equal to any rectilinear figure, it is worth inquiring whether it is not possible to prove that a rectilinear figure is equal to a circular area.

SB 2.G1

By *Elements* II, 14 any rectilinear figure can be transformed into an equal square; so the problem would be solved if *any* figure of equal area could be found, so long as it had straight sides. Notice that Proclus offered this account only as his opinion (writing, of course, many centuries later), and there is no historical tradition of how the problem came to prominence. But it seems that by the later fifth century BC its challenging nature was recognised. Not only was the problem notoriously difficult, but attempts to solve it usefully exposed the kind of problems to be faced in any geometric investigation of areas bounded by curves.

The earliest approach of which we have some record illustrates the problems well. This was the argument of Antiphon, an Athenian of the later fifth century. We know that he had some interest in the matter through a strange allusive remark by Aristotle, a century or so later:

Antiphon was a Sophist (a professional teacher of young men) like his contemporaries Protagoras and Hippias.

> It is for the geometer to expose the quadrature by means of segments, but it is not the business of the geometer to refute the argument of Antiphon.

SB 2.G2(a)

Fortunately, Aristotle's later Hellenistic commentators were more forthcoming. Themistius (fourth century AD) wrote that Antiphon

> inscribed an equilateral triangle in the circle, and on each of the sides set up another triangle, an isosceles triangle with its vertex on the circumference of the circle, and continued this process, thinking that at some time he would make the side of the last triangle, although a straight line, coincide with the circumference.

SB 2.G2(b)

Question 4 Draw a picture of Antiphon's construction, following the description given by Themistius. What do you take Antiphon's argument to have been, assuming that he claimed to have squared the circle by this means?

Comment

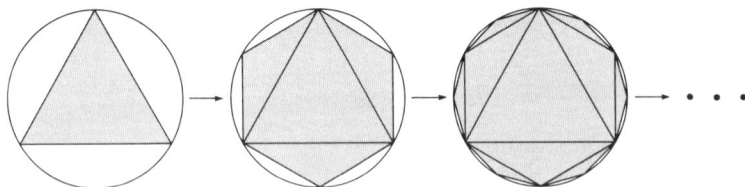

Figure 6

Antiphon's argument seems to have been that at some stage a polygon would have been constructed with the same area as the circle since its sides would coincide with the circle. Thus, since any polygon (a rectilinear figure) can be squared, so too can the circle. ■

Now it is clear from what Aristotle said that some flaw was seen in this argument. But it is not clear just what the problem was. Of course no matter how often the process is repeated, the straight sides of the polygon are *never* going to coincide with the curved circumference of the circle; that is, there is no number n for which at the nth repetition of the process the polygon and the circle will coincide. (Although of course if we try to draw it physically, the two will soon become indistinguishable, no matter how accurate our drawing instruments.) So it is tempting to read some such objection back to the time of Antiphon or shortly after. But this would be a geometrical objection, and Aristotle apparently saw the problem as lying outside geometry; so it could be Antiphon's concept of the circle itself to which Aristotle took exception. If Antiphon actually regarded the circle as essentially a polygon with a very large number of sides, then his claim that the circle could be squared would be quite correct; it just would not have anything to do with geometry—that is, it would lie outside the realm of investigations governed by the definitions used by other mathematicians. It could be an argument within an alternative mathematical development, a pre-Platonic tradition, drawing more strongly from appeals to the physical world and the evidence of the senses.

This is not the only interpretation of Antiphon's argument, nor of what was wrong with it. The commentator Alexander (second century AD) took the problem to be a violation of the principle that a circle can touch a straight line in only one point, or can cut it in only two and not more. However, Simplicius (sixth century) reported Aristotle's pupil Eudemus as seeing the problem as being that the remaining space between the ever-encroaching polygon and the circle would never be completely exhausted. There would always be some residual space, for otherwise 'the geometrical principle which states that magnitudes are divisible *ad infinitum*' would be violated.

In spite of all this heavy battery of sophisticated counterargument, there is something very appealing about Antiphon's construction. Whatever the principles violated, it is evidently a gallant attempt at getting somewhere near an answer. It is intuitively going in the right direction, even if formally wrong. Indeed, the principle of this approach—fill the space with bits whose areas you know—lies at the foundations of later Greek approaches to the problem, both of the circle and of curvilinear quadrature in general.

There was one other worker on the problem who had a good idea without the means to make it rigorous enough for Aristotle. The Sophist teacher Bryson, probably a contemporary of Plato, tried to square the circle by some argument which attracted Aristotle's scorn:

> The method by which Bryson tried to square the circle, were it ever so much squared thereby, is yet made sophistical by the fact that it has no relation to the matter in hand. . . . Bryson's argument is directed to the mass of people who do not know what is possible and what impossible in each department, for it will fit any. And the same is true of Antiphon's quadrature.

There are varying accounts by later commentators of what Bryson's argument was, but all agree that his construction put squares or polygons both inside the circle (as did Antiphon) *and also outside*. So his argument may have been the simple and rather unhelpful observation that the area of the circle is that of a square intermediate between the inscribed and the circumscribed square. (This is unhelpful, because finding that intermediate square is precisely the problem!) It has been suggested by the historian A. Wasserstein, however, that Bryson was seeking to *reduce* the problem to one of proportions, as Hippocrates had done for doubling the cube:

> I conclude that Bryson may well have hoped to build the quadrature of the circle into the framework of the theory of proportions. He may have suggested that there exists a sort of proportion in which what we now call π was the middle term between the circumscribed and inscribed figures as extreme terms, and that one could solve the problem of the quadrature of the circle by finding this proportion. His answer would then be of exactly the same nature and form as that of Hippocrates of Chios to the problem of doubling the cube. Like Hippocrates, Bryson would suggest that his problem could be reduced to another; and that that other problem belonged to the field of the theory of proportion. That he was not successful in this does not matter. From his point of view this must have looked a very promising line of attack. Bryson would then turn out to have been not a link between Antiphon, Eudoxus and Archimedes but an original thinker of the first rank.

Other historians take Bryson to have been observing that the circle is greater than any inscribed polygon, and less than any circumscribed polygon. This could lead to

The argument of Protagoras, about whether a tangent touches a circle at many points or just one (Section 4.1), hinged similarly on whether the senses or reason are the better guide to truth.

SB 2.G2(c)

SB 2.G3(a)

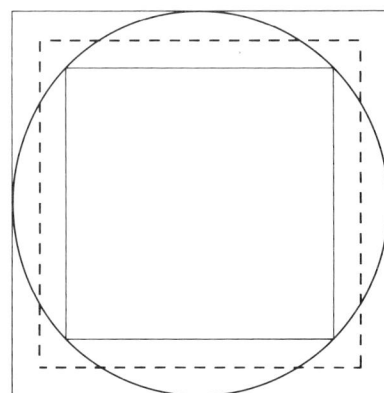

Figure 7

A. Wasserstein 'Some early Greek attempts to square the circle', *Phronesis*, **4** (1959), p. 100.

an argument that *if* the inner and outer polygons approach each other so closely as to coincide, then the resultant polygon (largest of the inner, smallest of the outer polygons) will constitute the squaring of the circle. As it stands, this argument does not explain how inner and outer polygons *can* coincide; but as with Antiphon the general idea, of 'compressing' the circle between inner and outer figures, would have been promising, and indeed turned out to be most powerful in the hands of Archimedes.

Question 5 Read Alexander's commentary on Bryson's approach, **SB** 2.G3(b). Which of the above views of historians do you consider is better supported by what Alexander said?

Comment ────────────────────────────────────

Alexander's account clearly states that Bryson put one square outside the circle, one square inside, and applied some argument in respect of a square in between them. He made no mention of further, Antiphon-like, constructions filling up the space inside the circle, or approaching it closer from the outside. This supports Wasserstein's view. ■

So perhaps Bryson was trying an alternative approach, but it did not lead anywhere. The more successful line of attack on the problem of curvilinear area proved to be a logical refinement of the approach pioneered by Antiphon.

What was the problem with arguments such as that of Antiphon? The point is this: suppose that we carry out the construction of inscribing polygons with more and more sides, getting closer and closer to filling up the circle—what then? At any particular stage of the process the circle has not actually been reached (still less some circumscribed polygon the other side of the circle). It is Antiphon's 'hand-waving' at this stage that perhaps caused all the bother. For instance, the possibility has not been ruled out that even if inner and outer polygons are constructed and each approach some ultimate area, they may not be the same; the largest inner polygon could differ from the smallest outer one. Again, there is no reason why the increasing polygons should ever fill up the whole circle, or indeed even some 'largest inner polygon' they seem to be approaching. Late fifth century Greeks (those interested in these matters, at least) were well aware of the difficulties in such arguments which seem to involve infinite processes. For instance, the Eleatic philosopher Zeno, a pupil of Parmenides, had put forward a series of paradoxes reported and criticised by Aristotle. One of these—the *dichotomy*—points out that in order to travel some distance one must have travelled half of it; then half of what remains (total distance = $\frac{3}{4}$); then half of what remains ($= \frac{7}{8}$); then . . . so to get there involves an infinite number of stages. Although this was probably not an insuperable problem to everyday Athenian journeying, it points up possible difficulties in arguments with an infinite number of steps—for example, on such an analysis it is not obvious that the destination is ever reached (or, in another form of the argument, that the journey can even begin).

Such difficulties with arguments such as Antiphon's are easier to put forward than to overcome. In evaluating this early work you should bear in mind its positive virtues as well as the logically unfinished arguments. It made it seem plausible that the circle—and curvilinear figures in general—*did have* an area property that could be investigated by some means. And other support for this belief was provided by Hippocrates' quadrature of lunes.

The mathematician credited with resolving the earlier difficulties, and placing curvilinear quadrature on a rigorous foundation, is Eudoxus (c.400–c.347 BC). A native of Cnidos (on the Ionian coast, just north of Rhodes), Eudoxus studied geometry under Plato's friend Archytas at Tarentum and in his early twenties came to study at the Academy in Athens. Later, in the 360s BC, he returned there (at about the time Aristotle came to the Academy as a young student) perhaps as something like a 'research fellow' in modern academic terminology. No work of Eudoxus is extant, unfortunately, and his achievements can only be inferred from what Archimedes and others said. But he seems to have been quite remarkably gifted, for besides his resolution of the quadrature problem (underlying Euclid's *Elements* Book XII), the proportion theory of *Elements* Book V reflects his ideas, and he was renowned too for his work in mathematical astronomy.

The essence of Eudoxus' technique for overcoming the above difficulties was twofold: a change in the attempted proof structure, and supplying a necessary axiom. To avoid the logical difficulties of the closing stages of arguments such as that of Antiphon he turned to an indirect argument, of the *reductio ad absurdum* form. For example, suppose that the quadrature of the circle C is sought and a rectilinear area A seems a likely candidate as the area to which inner and outer polygons seem to be approaching. Then, Eudoxus would have argued, *either* C does equal A *or* it does not; and if the latter can be shown to lead to a contradiction, then the former must hold and the quadrature of the circle has been found. That is the structure of the argument, which hinges therefore on being able to show that adopting the second alternative leads to an absurdity or contradiction. This is where the other aspect of his method, the new axiom, was needed. But let us now look at an example, both to get the structural aspect clear and also to identify the new axiom. We stay with the circle and look more closely at the proposition by Archimedes you studied briefly in the previous unit.

Question 6   Re-read Archimedes *Measurement of a Circle* Proposition 1 (**SB** 4.A1).

(i)   Describe the overall structure of the proof, to satisfy yourself that it fulfils the broad structural description above.

(ii)   Concentrating on the section marked 'I' ('If possible, let the circle be greater than K'), find the contradiction that enables him to show that the assumption is false.

(iii)   (*Harder*)   Start reading the proof from the beginning. Bearing in mind the kinds of logical difficulty which arose earlier, can you identify the point at which a similar problem seems liable to occur? That is, find a stage where you want to say to Archimedes 'Hang on! How do you know? What is your justification for that?'

Comment ───

(i)   Starting from the circle and from the triangle K, he says that *either* the circle equals K, *or* it is greater than K, *or* it is less than K. Showing that the last two possibilities lead to contradictions (in 'I' and 'II', respectively) enables Archimedes to claim that the first must hold.

(ii)   In 'I' he pursues an Antiphon-type construction, starting from a square and building up polygons with more and more sides, progressively filling up the circle. On the assumption that the circle is greater than K, he can obtain a polygon whose area lies in between that of K and the circle—larger than K, smaller than the circle. He then points out that this polygon is made up of little triangles whose height is less than the radius of the circle (being strictly inside it) and whose total perimeter is less than the circumference of the circle (because a straight line is shorter than a curved line between any two points). So, on either count, the area of the polygon must be less than K—which contradicts the previous assumption that it is bigger than K.

(iii)   You might have been alerted by the words 'and so on', at the point where the construction of inscribed polygons is described; this rather smacks of the 'hand-waving' that Antiphon seemed to go in for. If you read that paragraph carefully ('Inscribe a square $ABCD$, . . .') you can see that it is quite a complicated condition amounting to the claim that it is possible to carry out this process and obtain a polygon whose area lies between that of K and the circle. To this it is an entirely reasonable response to say: How do you know?   ■

Archimedes knew, though he did not spell it out here, because of an axiom of Eudoxus about how magnitudes behave. We do not know just how Eudoxus phrased it, but it was something along the lines of (in Aristotle's words):

> if I add continually [the same magnitude] to a limited magnitude, I shall at length exceed any assigned magnitude whatever, and if I continually subtract from it, I shall similarly make it fall short of any assigned magnitude.

Aristotle, *Physics* VIII in Heath, *Mathematics in Aristotle*, p. 153.

or (in Archimedes' words) in a more precise and slightly different form:

> Of unequal lines, unequal surfaces, and unequal solids, the greater exceeds the less by such a magnitude as, when added to itself, can be made to exceed any assigned magnitude among those which are comparable with it and with one another.

Archimedes, *On the Sphere and Cylinder*, Assumption 5.

or in a different form again, as a theorem in Euclid's *Elements* (X, 1):

> Two unequal magnitudes being set out, if from the greater there be subtracted a magnitude greater than its half, and from that which is left a magnitude greater than its half, and if this process be repeated continually, there will be left some magnitude which will be less than the lesser magnitude set out.

These three formulations differ in substantive and significant ways, which we do not explore further here. But what is in little doubt is that Eudoxus provided the logical linchpin on which the determination of complicated quadratures, and the solid (three-dimensional) equivalent *cubature*, could proceed. This is not unconnected with his theory of magnitudes presented in *Elements* V, whose significance we now briefly consider.

The above three versions of Eudoxus' idea are elaborations of what appears as Definition 4 of *Elements* Book V:

> Magnitudes are said to have a ratio to one another which are capable, when multiplied, of exceeding one another.

At first sight this may look innocuous or banal; in fact, it is quite momentous. It has two highly significant consequences.

First, it gives a criterion for when things can sensibly be compared, and so it effectively splits the rather broad concept of 'magnitudes' into categories of like things: areas, lines, volumes, forces, motions, for example. So in this way one area can be compared with another area, or one line with another line; but an area can have no ratio to a line, nor a volume to a motion, and so on. In other words, it spells out clearly that you can only compare comparable things. Secondly, things which do not behave like this are excluded altogether; in particular, the infinitely small and the infinitely large do not make an appearance, thus going far towards ruling out a worrying and paradoxical aspect of earlier mathematics.

Thus what may at first have appeared as tiresome or long-winded, namely the language of ratio and proportion in which so much Greek mathematics was couched, makes good sense. It was generally used, in geometry at least, until comparatively recently. Through the work of Eudoxus it also made possible the strong logical development that fostered the triumphs of Greek geometry in the next century and the remainder of this unit.

The overall pattern and style which you have seen exemplified in Archimedes' quadrature of the circle consisted of three aspects:

(i) inscribing and/or circumscribing known figures;

(ii) use of the axiom of Eudoxus in some form, to keep the behaviour of the magnitudes logically defensible;

(iii) all in a *reductio* proof structure.

Since the seventeenth century this has been called the *method of exhaustion*. We shall use this name too, as a handy label, but you should bear in mind that it is misleading in two ways. Firstly, it would be more accurate to call it a method of *non*-exhaustion, if anything; the point is to *avoid* the hand-waving tangle in which Antiphon seems to have landed through a direct exhaustion approach of trying to fill up the space completely.

Secondly, it is not a *method*, if by that one understands a coherent theoretical device, a set of rules for generating the answer to any quadrature or cubature problem. It is more of a broad approach or loose set of guidelines wrought through practice. Any particular problem had to be thought through and crafted afresh, albeit with the experience of similar problems to help. This is rather characteristic of the Greek mathematical style; we just do not find the general algorithmic kind of approach that you will be able to identify in some later mathematical styles, such as the development of algebra and of the calculus.

The word 'of' in 'method of exhaustion' is unexceptionable, though.

A historiographical warning to end with. What you have learnt in this section has been a simple and straightforward story about rather difficult subject matter. What we might be able to say were the historical record more complete would surely be more complicated. There is nothing like losing almost all one's sources for ensuring a nice rational storyline.

12

4.3 ARCHIMEDES

You may have noticed earlier that, whenever we needed an interesting or significant example, we were drawn towards the work of Archimedes. The reason for that is easily explained—he was the most creative and prolific of those mathematicians whose works have in large measure survived. As with Euclid's *Elements*, this survival was no accident, but a testimony to his influence and the high regard in which he was held by succeeding generations. No one we have met so far has shown such astonishingly broad interests, and excelled in so many fields. In this section we survey his life, death and other achievements, in order to broaden your perspective on the richness of Greek mathematical activities, before discussing his contributions to the Greek study of curves.

Born around 287 BC, Archimedes seems to have spent most of his life in his home town of Syracuse (in Sicily). He may have spent some time in Alexandria, perhaps studying mathematics there a generation or two after Euclid; he certainly had friends there.

Question 7 Several of Archimedes' works come down to us with the covering letters he wrote, a revealing source for learning about the organisation of mathematical research in the third century BC. Read these letters (**SB** 4.A3, 5, 7, 8 and 9), and see what impression you can form about this.

Comment

He had two correspondents in these letters. Most of the letters were to Dositheus, whom Archimedes does not seem to have known personally, but whom he adopted as a correspondent after the death of their mutual friend Conon; and one letter was to Eratosthenes, who was Librarian at Alexandria. Archimedes seems to have used these people as channels to communicate with 'persons engaged in mathematical studies', setting them problems and subsequently supplying the proofs if the answers were not forthcoming (as was generally the case).

The impression I get, reading between the lines, is of a few groups or individuals, perhaps dotted around the Mediterranean, keeping in touch by letter on what was happening at the forefront of geometrical research. Note that it was quite advanced material which was under discussion, and that Archimedes clearly saw himself as in a historical tradition: he spoke of past mathematicians, including 'all the many able geometers who lived before Eudoxus', and also of the advances that could be made by his contemporaries or by 'my successors'. It could be that much of the mathematical activity at the time of Archimedes took place in Alexandria, or else that Alexandria was more of a 'clearing-house' for communication to and from more scattered researchers. ■

Archimedes was interested in all the mathematical sciences (except possibly music): astronomy, mechanics, statics (levers and centres of gravity), and hydrostatics (bodies floating in water). What is especially noticeable by comparison with many of the earlier mathematicians is the way he combined the two streams of activity which Plato had been so keen to separate—pure geometrical analysis, and the mechanical or practical. Thus his work on the lever is purely geometrical, with propositions derived logically from postulates in a Euclidean-looking way. But his astronomical book *On Sphere-making* (now lost) was about constructing a planetarium modelling the motions of heavenly bodies. Indeed, his reputation in the centuries after his death was rather more as a maker of mechanical marvels than as a geometrician. His planetarium was still extant two centuries later, and was seen by the Roman statesman and author Cicero (106–43 BC), who wrote:

> When Gallus set the sphere in motion, one could, at every turn, see the moon rise above the earth's horizon after the sun, just as occurs in the sky every day; and then one saw how the sun disappeared and how the moon entered into the shadow-cone of the earth with the sun on the opposite side....

Cicero, 'De republica', quoted in B. L. van der Waerden, *Science Awakening* (Oxford 1961) p. 211.

It would be interesting to know how Archimedes himself regarded his mechanical inventions. That would help us to understand whether, from our perspective, he was essentially a 'pure' mathematician working within the geometrical research tradition, who happened to have mechanical interests as a sideline, or whether he was a more complex person who defies categorisation and was consciously working within a broader context. We have some evidence on this point in one of the most interesting accounts of Archimedes, by the later Greek historian Plutarch.

Plutarch lived from about AD 46–120 in Boeotia (in mainland Greece, by then part of the Roman Empire). Having studied Platonic philosophy at Athens, he devoted himself mainly to a life of learning. The *Lives* he wrote of great Greeks and Romans are an important historical source because so many of the sources he drew upon are now lost. However, Plutarch's interest in biography was primarily ethical, in moral character, and so his attentiveness to historical detail and accuracy is not necessarily that of a historian. His account of Archimedes comes in his life of the Roman general Marcellus, head of the Roman troops whose siege of Syracuse (214–212 BC) ended in its capture and the death of Archimedes.

Question 8   Read Plutarch's account (**SB** 4.B1) to try to ascertain Archimedes' attitude towards the question raised earlier, of the relative values he placed on his geometrical and mechanical activities. You might find it helpful to pursue the following questions.

(i)   What does Plutarch say Archimedes' attitude was?

(ii)   Is any evidence given or implied for this assertion?

(iii)   How reliable, overall, does Plutarch seem to be about Archimedes? Does his general account seem plausible and taken from some first-hand or eye-witness report?

(iv)   In view of what you know about Plutarch (from the above), in which direction might his account of Archimedes be biased, if at all?

(v)   Can you conclude anything (either about Archimedes' view, or more generally) about attitudes towards geometry and mechanics?

Comment ───

(i)   'He regarded the business of engineering, and indeed of every art which ministers to the material needs of life, as an ignoble and sordid activity, and he concentrated his ambition exclusively upon those speculations whose beauty and subtlety are untainted by the claims of necessity.'

(ii)   Hardly any, although the observation that he wrote little about his mechanical discoveries is a telling supporting point; and his choice of tomb inscription (at the end of that paragraph) can reasonably be taken as showing what Archimedes considered to be his most important achievement.

(iii)   Plutarch's account of the role of Archimedes' inventions in the defence of Syracuse reads as though it had grown somewhat in the telling, a kind of formerday Baron Munchausen. Notice that he offers three versions of the death of Archimedes—and that all three are subservient to what Plutarch is *really* interested in: the response to his death by Marcellus, the noble victor. So Plutarch gives us no great confidence that he had access to a reliable primary source, but rather that he is regaling us with the legends that grew up around the events. Plutarch is not unaware of the dangers—his saying 'it is not at all difficult to credit some of the stories which have been told about [Archimedes]' implies that he would be prepared to criticise any story too implausible. But this leaves us in some doubt about how reliable Plutarch is about Archimedes' views.

(iv)   It is important to bear in mind that Plutarch was a Platonist. He attributed to Plato (second paragraph) strong views about Eudoxus and Archytas having 'corrupted and destroyed the ideal purity of geometry' through invoking mechanical instruments in geometry. (This is not inconsistent with Plato's views, although this episode does not appear in the extant Platonic writings.) Consciously or otherwise, Plutarch would have wanted to see the great mathematician Archimedes as following the precepts of Plato. So Archimedes' traditional enthusiasm for mechanical devices could be something that, for Plutarch, needed explaining away.

(v)   The implication of (iii) and (iv), taken together, is that it would not be safe to draw any conclusion from Plutarch about Archimedes' views on the question (though the evidence cited in (ii) is worth taking into account). More generally, however, this account does offer first-hand evidence that the status of geometry versus mechanics was still a live issue in the first century AD, and that the views of Plato continued to be influential in the living academic tradition after nearly half a millennium. ■

There are a number of other interesting perceptions in what Plutarch wrote, to which reference will be made when appropriate. We turn now, however, to aspects of Archimedes' mathematical work. We can highlight only a few significant aspects of this, described by Sir Thomas Heath as 'a sum of mathematical achievement unsurpassed by any one man in the world's history'. This is a strong claim, but see what you think.

T. L. Heath, *A History of Greek Mathematics*, vol. II (Dover, 1981) p. 20.

First, let us consolidate the 'squaring the circle' story. In the previous section you looked at Proposition 1 of *Measurement of a Circle*, where Archimedes proved that the circle is equal to the right-angled triangle whose sides are its radius and its circumference. There we were interested in the proof structure; now let us look at the significance of the result.

1 Recognise carefully that this result does not 'square the circle' in the terms of the traditional problem. Archimedes showed that the circle does indeed have a squarable area, but the problem was to *construct* that area by line-and-circle construction. Archimedes reduced the problem, in effect, to that of constructing the circle's circumference as a straight line. Also, as you saw in Section 4.1, he reduced the problem further by showing, in *On Spirals*, that the circumference of the circle is rectifiable (that is, can be constructed as a straight line) if you can construct a tangent to the spiral. This is not possible by line-and-circle construction, however.

2 This proposition was remarkable, for the best previous result on the matter had been what appears as Euclid's *Elements* XII, 2: 'Circles are to one another as the squares on the diameters'. That is the result which Hippocrates used in his quadrature of lunes a century or more earlier, though presumably he could not have proved it rigorously since the Euclidean proof (**SB** 3.E3) is by the method of exhaustion developed, we believe, by Eudoxus.

The significant development marked by Archimedes' proposition is that for the first time (that we know of) two constants connected with measuring circles were shown, in effect, to be the same. The two constants involved are

As we now call both these constants 'π', this is slightly confusing to explain!

(i) the ratio of the area of the circle to the square on the radius (the π of $A = \pi r^2$);

(ii) the ratio of the circumference of the circle to its diameter (the π of $C = 2\pi r$).

It is by no means obviously true that the first one of these *is* a constant, the same ratio for all circles, but this seems to have been recognised implicitly from quite early times, in both Egypt and Mesopotamia. It was first proved, in effect, in *Elements* XII, 2. But that the two πs are the same—again, far from obvious, though our modern symbolism entirely disguises the fact—was not made apparent until the work of Archimedes.

3 Also rather striking, and no less influential, was Proposition 3 of *Measurement of a Circle*, where Archimedes established bounds for this ratio:

> The circumference of any circle is greater than three times the diameter and the excess is less than a seventh part of the diameter but more than ten seventy-firsts.

In symbols, $3\frac{10}{71} < \pi < 3\frac{1}{7}$.

He did this by inscribing and circumscribing polygons of 96 sides, and basing the bounds on his calculations of their areas. The number 96 arises by starting with a hexagon (6 sides) and progressively doubling the number of sides so as to get polygons of 12, 24, 48 and 96 sides, so becoming ever closer to the circle. By taking this doubling process further one could get bounds as narrow as one wants, but even in the case considered by Archimedes the computations demanded high number-handling skills. Unfortunately the text that has come down to us lacks the full details of the calculations, but it is clear that Archimedes was a very skilled numerical calculator. This is something one would not expect, to judge by the earlier products of the Greek geometric tradition, in which the very notion of using numbers to calculate with—still less to approximate with—was rather foreign. So Proposition 3 is evidence that Archimedes was not working wholly within the strict geometrical research tradition, and that the computational tradition and skills evident in some later Hellenistic mathematics (notably in the work of Hero of Alexandria, and in the mathematical astronomy culminating with Ptolemy in the second century AD) date back to at least the time of Archimedes. Whether there was a significant Greek computational tradition even earlier, cast into historical shadow by the glare of Platonism, or whether this speaks of influences from Babylon or elsewhere in the Middle East, is not really clear at present.

Archimedes seems to have rather enjoyed numbers. He wrote two works on ways of expressing large numbers. There was no very ready means of doing this in everyday Greek numerals (unlike, say, our place-value system which is capable of indefinite extension), and in a work called *The Sand-Reckoner* Archimedes described how to form very large numbers. The basic idea was to count as far as one could (in Greek that would be a myriad myriad ($= 100\,000\,000$)), then use *that* as a unit for the next stage, and so on until a myriad myriad to the power myriad myriad was reached; then use *that* as a fresh unit; and so on. In this way he reached a number which in our notation would have eighty thousand million million ciphers, that is 1 followed by $80\,000\,000\,000\,000\,000$ zeros. He then proceeded to show that this was sufficient to count the number of grains of sand that would fill the universe, even the universe as large as that required by the hypothesis of Aristarchus.

There is a further point worth making about Archimedes' *Sand-Reckoner*. In much Greek cosmology, Aristotle's in particular, the heavens were constituted differently from the earth, and different laws applied (albeit ones which could be described in terms of pure geometry). To break from this assumption was one of the great seventeenth-century developments, as you will see in *Unit 7*. So for Archimedes to conceive the very earthy image of the whole cosmos filled with sand, even in a spirit of exuberant fantasy, is further testimony to his unfettered free-thinking approach.

So this work by Archimedes is of further great historical significance, as the place where we learn of the views of Aristarchus on a heliocentric universe, which Copernicus was aware of in devising his not dissimilar system in the sixteenth century. See Archimedes' preface, **SB** 4.A2.

In quadrature and cubature, Archimedes' work was within the geometrical tradition and marks its Greek culmination. Nothing so powerful was to be seen again for nearly two thousand years. It is difficult to convey this convincingly, however, without a much more detailed study. It seems paradoxical that what is generally held to be the arena of his greatest achievements—a claim he would have agreed with, to judge by the inscription he specified for his tomb—is one where Archimedes' contribution is not visibly different on a first glance from the work of his predecessors. It is just the same kind of thing as one finds in Euclid's *Elements* XII (areas and volumes of this, that and the other), only more so. He breaks no radically new ground, here, either in subject-matter or in approach as seen in the finished proof. But a closer examination reveals that the problems are intrinsically very difficult, and Archimedes' handling of them is flexible and versatile. Plutarch must have been referring to books such as *On the Sphere and Cylinder* and *On Conoids and Spheroids* when he said:

> Certainly in the whole science of geometry it is impossible to find more difficult and intricate problems handled in simpler and purer terms than in his works.

Examples of what Archimedes proved in the former work are

> The surface of any sphere is equal to four times the greatest circle in it.

and

> Any sphere is equal to four times the cone which has its base equal to the greatest circle in the sphere and its height equal to the radius of the sphere.

which has as a corollary the great result engraved on his tombstone, that a cylinder is one-and-a-half times the sphere it just contains.

Archimedes, *On the Sphere and Cylinder*, Book I, Propositions 33-34. In modern arithmetical terms this is to say that the surface area of a sphere of radius r is $4\pi r^2$.

The volume of a sphere of radius r is $\frac{4}{3}\pi r^3$, and that of its circumscribing cylinder is $2\pi r^3$, which is one-and-a-half times $\frac{4}{3}\pi r^3$.

We now look in more detail at one of the results in *On Conoids and Spheroids*, as our main example of what an Archimedean proof looks like. Conoids and spheroids are surfaces of revolution of the conic sections—just as a sphere can be thought of as a circle revolved round its diameter, so too surfaces arise by revolving other conic sections (ellipse, hyperbola, and parabola) around their diameters or axes (Figure 8).

Proposition 21 of *On Conoids and Spheroids* shows that the (volume of a) right-angled conoid (a paraboloid of revolution, in modern terms) is one-and-a-half times the cone on the same base and axis. Like the results mentioned above, the proof is by the method of exhaustion.

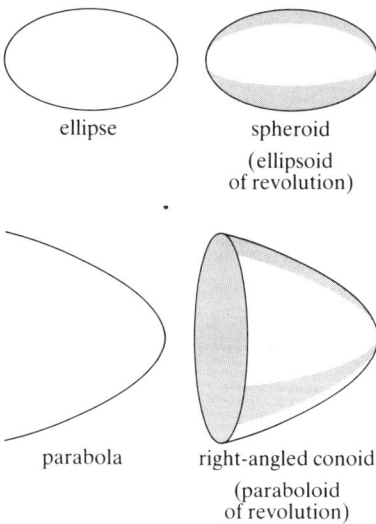

$$r : s = p^2 : q^2$$

ellipse

spheroid
(ellipsoid
of revolution)

ordinates

q p

r

axis s

parabola

Figure 9

parabola

right-angled conoid
(paraboloid
of revolution)

Figure 8

Question 9   Read through Proposition 21 of *On Conoids and Spheroids* (SB 4.A8) in order to satisfy yourself that the proof has the three characterising features of the method of exhaustion, as given at the end of Section 4.2.

Study Note   Do not feel you must work through the proof in detail, though that would be informative. We shall be looking at the properties of a parabola in more detail in the next section; here you just need to know that distances along the axis are as the squares on the ordinates—see Figure 9.

Comment ───

Do not worry if you found this hard to sort out—part of the complexity arises from the fact that Archimedes is dealing with the problem in a very general way. But you should have been able to see that, after describing the construction and labelling of the paraboloid (of which a section, a slice down the axis, is shown in the diagram), he produced a cone X and moved into a *reductio* proof to show that the paraboloid is equal to X. ('If not, the segment must be either greater or less than X. . . .')

In doing this, Archimedes fitted cylindrical discs, inside and outside the surface, so that the paraboloid is roughly approximated by two 'step pyramid' formations. The volume of each little cylinder is known, so the volume when they are all piled up is known too.

The third crucial aspect of the method of exhaustion was the axiom of Eudoxus in some form. This does not appear explicitly here, but we can infer that it was invoked in the previous supporting proposition. The revealing sentence is:

> We can then inscribe and circumscribe, as in the last proposition, figures made up of cylinders or frustra of cylinders with equal height and such that
>
> (circumscribed figure) − (inscribed figure) < (segment) − X.

For this is saying that *whatever* the difference between the paraboloid and the cone X, one can construct cylindrical discs so that the inner and outer step pyramids are closer together than that difference. It is this condition that tells you the magnitudes are behaving as they ought—the rest of the proof is just spelling out the consequences. ■

So this proof is by the method of exhaustion and, from the perspective of logical rigour, nothing could be more satisfying, nor a greater tribute to the sophistication of the Greek geometric research tradition. But there is one problem outstanding. How did Archimedes know that this was the result he wanted to prove? How did he

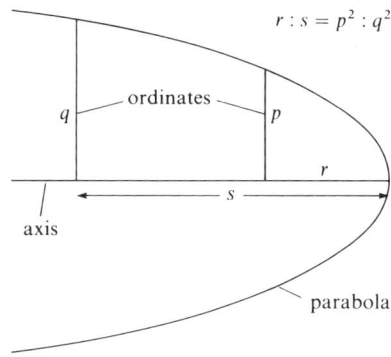

know, *in advance*, that the paraboloid is one-and-a-half times some cone? Why is it not twice, or $\frac{81}{64}$ times, or something which is not a commensurable ratio at all?

It is characteristic of a *reductio* proof that the result to be proved must be known in advance, so this problem occurs with all exhaustion proofs. Brilliant though Eudoxus' method was for enabling logically rigorous proofs to be formulated, and proving beyond doubt that some result was the case, it made no helpful contribution to finding out the result in the first place. Once mathematicians in the seventeenth century began to study the works of Archimedes with care, this struck them as puzzling. The suspicion grew that the method of analysing a problem so as to discover the answer before embarking upon its synthetic proof had been more developed in Greek times than the few scattered references suggested. The English mathematician John Wallis wrote in 1685 that Archimedes seemed

> as it were of set purpose to have covered up the traces of his investigation, as if he had grudged posterity the secret of his method of inquiry, while he wished to extort from them assent to his results.

and that

> not only Archimedes but nearly all the ancients so hid from posterity their method of Analysis (though it is clear that they had one) that more modern mathematicians found it easier to invent a new Analysis than to seek out the old.

J. Wallis, *Treatise of Algebra*, cited in Heath, *A History of Greek Mathematics*, vol. II, p. 21.

There matters would have stayed were it not for a remarkable find early this century.

In 1906 J. L. Heiberg discovered a parchment in Constantinople from which various Archimedean works could be recovered. Most notably, there were traces of a hitherto lost work called *On Mechanical Theorems, Method to Eratosthenes* (or *The Method*, for short). The copy had been made in handwriting of the tenth century, and at some later time partially scraped off, to be overwritten by a religious text. This work is of considerable interest, in helping to resolve the problem of how Archimedes reached some of his results.

Heiberg was the Danish scholar whose critical editions of Greek mathematical texts form the foundation of modern historical scholarship on the period.

Question 10   Read the introductory letter of *The Method* (**SB** 4.A9(a))   Does it confirm or rebut Wallis' conjecture quoted above?

This course of events is a further reminder of how slender and chancy the thread can be that leads us to our knowledge of the past. It also encourages us to hope that there may still be important Greek texts that will resurface one day.

Comment ————————————————————————

A little of both. Archimedes did indeed have a method for discovering results, so the seventeenth-century suspicion was quite correct. On the other hand, Wallis' aspersions on Archimedes, that he 'grudged posterity the secret', is quite wrong since Archimedes sent the method to Eratosthenes hoping to make it available to other mathematicians. He was not responsible for its having been largely overlooked thereafter. ∎

We can best see what Archimedes' method was by looking at an example. Proposition 4 is the result you have just studied on the paraboloid of revolution, so let us examine that. Please *read it now* (**SB** 4.A9(c)), and then follow the summary.

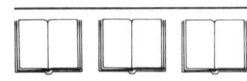

Start from the diagram on **SB** p. 171. It is a vertical slice through the middle of something with four components: the paraboloid in question ($BOAPC$), a cone BAC inside it, a cylinder $BEFC$ outside it, and a bar DAH going along the common axis, with A as its midpoint. Note that the paraboloid, cone and cylinder all have a circular base with diameter BC. We are to think of the bar as that of a balance. Horizontal balance bars may be more familiar so we have rotated the diagram (Figure 10).

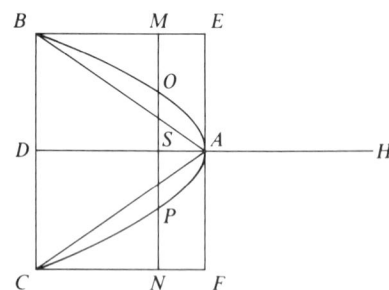

Figure 10

We now consider just the paraboloid and the cylinder (Figure 11). The upshot of the method is that Archimedes took the paraboloid away, hung it from the bar at H, and observed that the paraboloid and the cylinder are then in balance. Since the weight of the cylinder acts through its centre, which is half-way along, the cylinder is twice the paraboloid (see Figure 12). The result then follows as the cylinder is, by a result in Euclid, three times the cone.

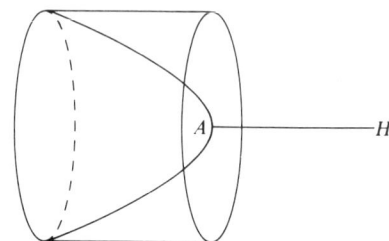

Figure 11

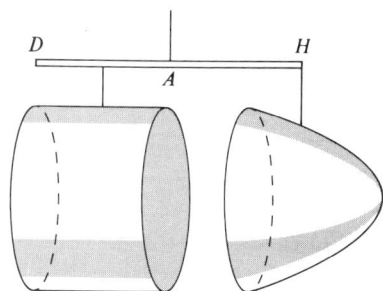

Figure 12

Figure 13

paraboloid slice
transferred to
here

slice

Having seen where we are heading, let us go back and fill in the middle part of the argument. Archimedes took any slice through the objects, parallel to the base. The slice produces *circular* cross-sections of both cylinder and paraboloid (Figure 13).

Archimedes observed that the circular cross-section of the paraboloid, if moved to the far end H of the bar, would balance the circular cross-section of the cylinder (left where it is). The reason for that is an argument in which the fact that we are dealing with a paraboloid (i.e. a rotated *parabola*) is critical. The property of the parabola needed is that mentioned earlier (Question 9), that distances along the axis are as squares on the ordinates; so $DA : SA = BD^2 : OS^2$. Or, since HA was designed to be equal to DA, $HA : SA = MS^2 : OS^2$. (Note that $MS = BD$, because that is the constant radius of the cylinder.) Now by *Elements* XII, 2, circles are as the squares on their diameters—or as the squares on the radii, which is what we have here—so

$HA : SA$ = circular cross-section of cylinder : that of the paraboloid.

At this stage, the equation is precisely that of the law of the lever. Archimedes proved in *On the Equilibrium of Planes* that

> Two magnitudes, whether commensurable or incommensurable, balance at
> distances reciprocally proportional to the magnitudes.   SB 4.A4

This is just what we have here. So the circular cross-section of the paraboloid (taken at distance HA) balances the circular cross-section of the cylinder taken where it is, at distance SA. Since this conclusion applies to any slice we care to make, the effect is that all slices of the paraboloid (that is, the whole paraboloid) can be piled up at H, and will balance the cylinder as it is. Re-read the proof in Archimedes' wording now, to consolidate your understanding.

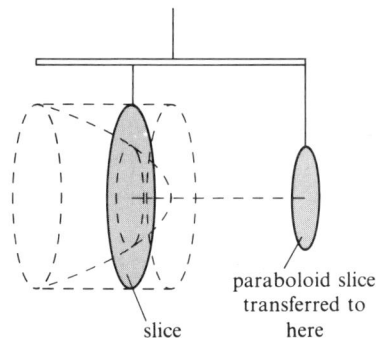

This proposition is a good example of Archimedes at his most versatile and flamboyant. If *The Method* was not generally taken up, that could be because no-one else could follow it so well. Plutarch remarked on the difficulty of discovering Archimedes' synthetic proofs beforehand, and his remarks seem equally applicable to *The Method*, despite the hopes Archimedes expressed in his letter to Eratosthenes that others could make discoveries through it.

Archimedes wrote to Eratosthenes that the results found through this method 'had to be demonstrated by geometry afterwards because their investigation by the said method did not furnish an actual demonstration'. Why is this? Why does the argument you have just been examining not count as a *proof* of the result? There were two difficulties, and historians have disagreed over which was uppermost in Archimedes' thoughts. The first of these is that, following both Plato and Aristotle, there could be unease about the intrusion of 'mechanical' concepts. The very notion of weights balancing is rather mundane and ungeometrical in the Platonic view, and Aristotle had emphasised that purely geometrical properties such as *area* could not be properly proved by reference to concepts such as *weight* or *centre of gravity* arising in a subsidiary science such as mechanics. The second difficulty is that the technique of slicing and moving the slices around appears to reawaken the logical difficulties we discussed earlier. For an elementary paradox appears very quickly, in conjunction with what Archimedes wanted the slices to effect. In order for the law of the lever to apply, the slices must have weight—zero magnitudes cannot enter the process; but to consider the cylinder and paraboloid as made up of all their slices, each slice must be of zero thickness and thus zero weight, for although the cylinder can be considered as made up of thin cylindrical discs, the paraboloid certainly cannot. Thus we are back to a solid analogue of the problem that may have earned Antiphon Aristotle's scorn—a paraboloid made up of circular slices is no *logical* advance on a circle made up of an indefinite number of straight lines. Or to put it another way, much the same heuristic intuitions as may have guided Antiphon are evident, in a more sophisticated framework, in the work of Archimedes 150 or more years later. And they were to be rediscovered, without the benefit of *The Method*, by mathematicians of the early seventeenth century.

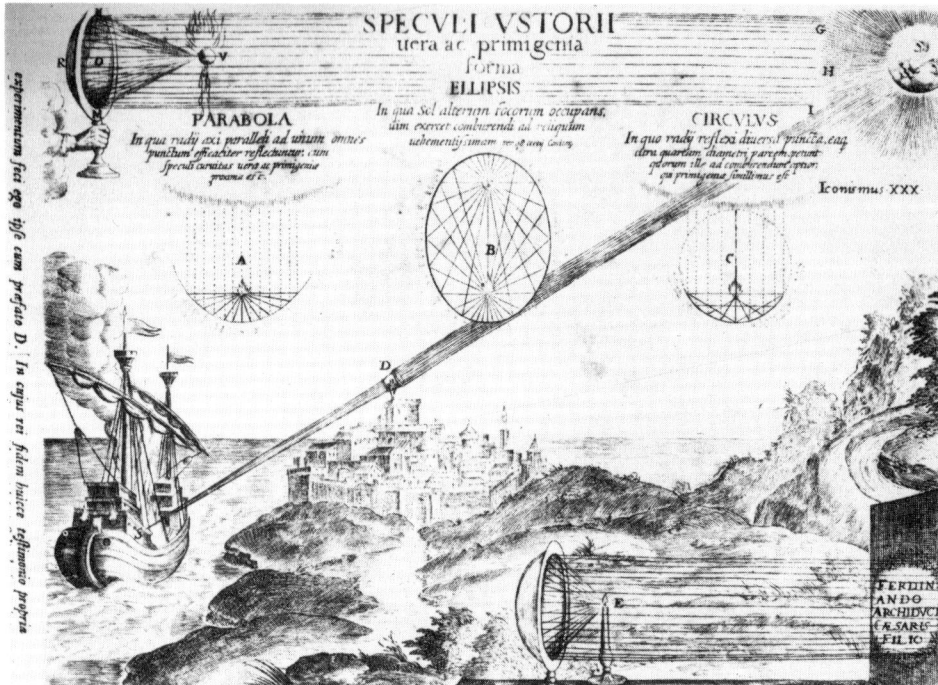

Figure 14 Fascination with Archimedes' life and reputed achievements has continued. This illustration from Athanasius Kircher's *Ars Magna Lucis* (1646) explores how Archimedes might have set fire to the Roman fleet by focussing the sun's rays. 'Burning mirrors' were certainly a subject of research in the later third century, as the work of Diocles (next section) shows.

4.4   THE GREEK STUDY OF CONICS

Although several different curves were known to the Greeks from various contexts, only one family of curves was studied in depth by many people: the conic sections. Discussion of these curves originally arose in connection with the problem of doubling the cube. The name of Menaechmus, 'a student of Eudoxus who also was associated with Plato', is traditionally given in connection with their discovery and early development, somewhere around 360–350 BC. The early history of conic sections is somewhat obscure, however, for much the same reason which makes the history of geometry before Euclid only possible through painstaking reconstruction. Towards the end of the next century a work appeared which became the definitive text, the *Conics* of Apollonius. Earlier treatises by Euclid and by a certain Aristaeus are mentioned by the commentator Pappus, but are now lost.

Proclus, **SB** 2.A1

We shall try nevertheless to see what happened between the times of Menaechmus and Apollonius. First, what was it that was discovered some time in the mid-fourth century? Note that the name 'conic section' both refers to a curve and reflects a particular way of defining that curve (by its genesis as a section of a cone).

Historically these two aspects may be distinct, however. It is generally accepted that it was in doubling the cube that mathematicians of the Academy constructed curves later perceived as sections of cones. But the perception of these curves *as* conic sections may indeed be from later in the century, and due perhaps to Aristaeus. The historian Wilbur Knorr has argued that:

The connection between conic sections and two mean proportionals was discussed in *Unit 3*, Box 4.

> Menaechmus and the geometers in the decades immediately before and after him initiated not a theory of the conic sections, but a body of geometric problems solved according to a form of the method of analysis. When the resolving loci turned out to be curves which were later known as conic sections, these were at first constructed, if at all, via point-wise procedures. Only late in the fourth century, near the time of Euclid, did one conceive of generating a class of curves via the sectioning of cones and begin their geometric investigation.

W. Knorr, 'Observations on the early history of the conics', *Centaurus*, **26** (1982) p. 7.

It was thus by the end of the fourth century, at the latest, that the conic sections were defined (by genesis) as sections of cones, the sections being made perpendicular to a side of the cone. Depending on the shape of the cone three different kinds of curve could be produced. If the angle at the top is less than a right angle, the section gives rise to what Apollonius later called an *ellipse*; if a right angle, to what he called a *parabola*; if greater than a right angle, to what he called a *hyperbola*.

We come to the significance of these names later, but for the moment we use them without explanation.

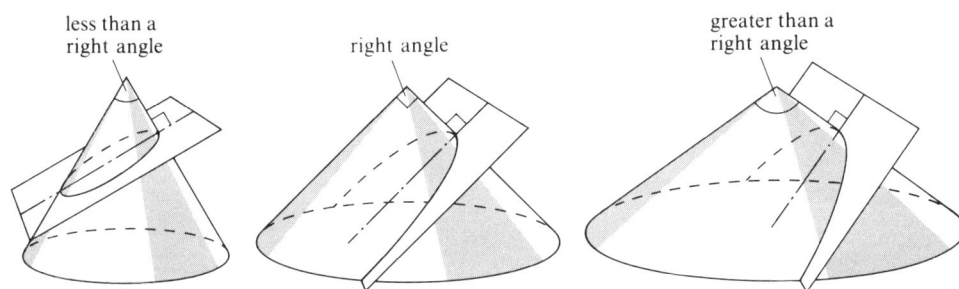

Figure 15

Of course, such a definition is of little use in itself unless further mathematical properties can be deduced. In fact from these definitions it is not a difficult application of elementary plane geometry to find their *symptom*: the condition which points on the curve satisfy. The symptom of the parabola is

$$(\text{ordinate})^2 = (\text{abscissa}) \cdot (\text{constant parameter})$$

as is shown in Box 3. Or, in the form we met the symptom in the work of Archimedes (Question 9), for any two points on the parabola,

abscissae are as the square on the ordinates.

By the time we meet conic sections in the work of Archimedes, they generally make their appearance directly through the symptom. That is, Archimedes worked from this condition, taking it to be something evidently known and not needing to be derived afresh, in such a way that the 'definition by genesis' involving cones had now in effect become a 'definition by property'. It is now the symptom of the curve that serves as its defining property.

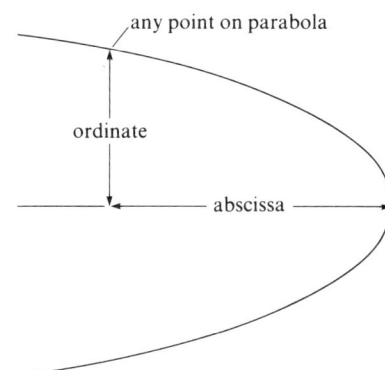

Figure 16

Before looking at the contribution of Apollonius himself, we move on to an interesting work contemporary with his, *On Burning Mirrors* by Diocles (early second century BC). As the last known discussion of conics written before the influence of Apollonius spread, this work is a guide to the developments reached earlier. Please *turn now* to the extracts in **SB 4.C**.

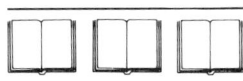

The introduction (**SB** 4.C1) describes the problem, which is associated with the familiar names of Archimedes' correspondents Conon and Dositheus, and is to do with reflections from mirror surfaces. The problem of finding a mirror to reflect the sun's rays to a single point was solved practically, Diocles says, by Dositheus.

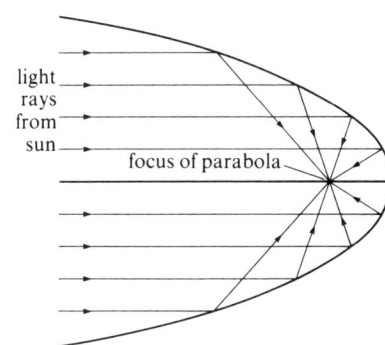

Figure 17

In Proposition 1 (**SB** 4.C2) Diocles proved this result by considering properties of the parabola. The property derived theoretically is that every ray coming in parallel to the axis of the parabola is reflected to the point D, which we now call the *focus* of the parabola.

The focus is easily related to the symptom of the parabola; its distance from the vertex B of the parabola is a quarter of the 'constant parameter', in the formulation of the symptom given above. It is worth noticing that for Diocles the parabola comes equipped, as it were, with a line BH called 'half the parameter of the squares on the ordinates'. This is half the constant parameter of the symptom, which is the distance from the original cutting plane to the point of the cone (AD in Box 3). We need not work through Diocles' proof in detail, though it is worth observing his geometric style.

Go through the argument in Box 3 if you are interested, but we shall not need it subsequently. If you find it taxing that is probably not because of each straight-forward geometrical argument but because of the difficulty of juggling mentally with three different intersecting planes.

Box 3   Note on deriving the symptom (property) of the parabola from its definition as a section of a right-angled cone.

We start with the cone, sectioned so as to produce the parabola.

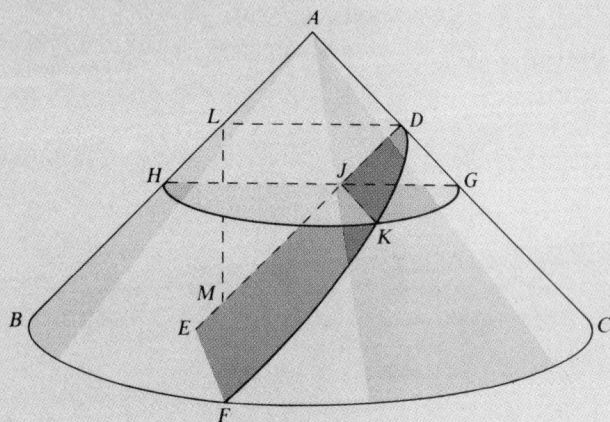

In the diagram, DKF is the front half of the parabola, coming forward out of the plane of the paper, and DE is its axis. So DE is at right angles to the cone's side AC, and is parallel to the cone's side AB. Now take any point K on the parabola. K is also, therefore, on the surface of the cone. Take a circular section HKG of the cone.

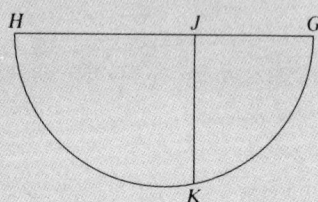

$$GJ : JK = JK : JH$$

Then $JK^2 = HJ \cdot JG$, because an ordinate in a semi-circle is a mean proportional between the lengths of diameter it cuts. But $HJ = LD$ because the parabola's axis is parallel to the side AB, and so $JK^2 = LD \cdot JG$. But the triangles LDM and JDG are similar—they have the same angles—so their sides are in the same ratios, that is $LD : DM = DJ : JG$, or $LD \cdot JG = DJ \cdot DM$.

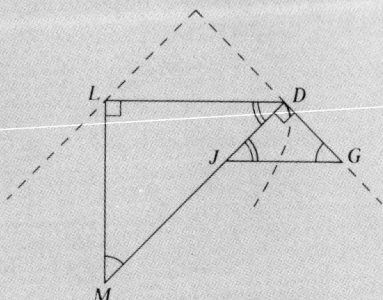

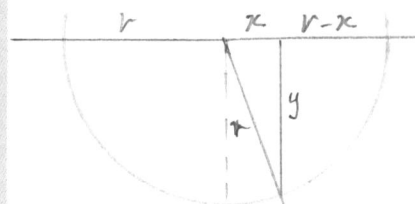

$$y = \sqrt{r^2 - x^2}$$

then

$$\frac{r-x}{y} = \frac{y}{r+x}$$

$$\implies y^2 = r^2 - x^2$$

$$\implies \sqrt{r^2 - x^2}^2 = r^2 - x^2$$

22

But *LDM* and *ADL* are similar right-angled isosceles triangles, such that the side *LD* of one is the hypotenuse of the other. It follows that $DM = 2AD$, so $JK^2 = DJ \cdot 2AD$ which is the required property of the parabola.

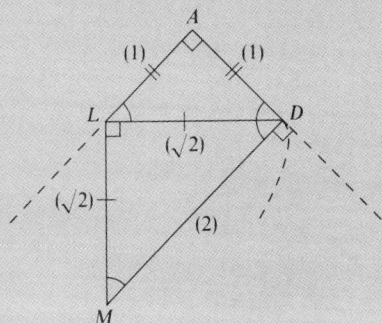

In words, any point on the parabola has the property that the square on its ordinate (the square of its distance from the axis of the parabola) equals the rectangle one of whose sides is the abscissa (how far down the axis the ordinate cuts) and whose other side is a constant for that parabola, namely twice the distance down the cone of the cutting plane. (Putting $JK = y$, $DJ = x$, $2AD = p$, we get $y^2 = px$, the familiar 'Cartesian equation' for the parabola.)

This reconstruction is due to the nineteenth-century German historian C. A. Bretschneider, and uses only theorems of plane geometry that were well known by the time of Menaechmus.

Question 11 Read over Diocles' proof (**SB** 4.C2) to form an impression of what level the handling of conic sections had reached by his time.

(i) Does he work through properties of the parabola from first principles, from the definition as a conic section or from the symptom?

(ii) Does Diocles' work seem to lie entirely within the geometric research tradition?

Comment ————————————————————————————————

(i) No, he does not even work from the symptom of the curve, as you could have perceived from the fact that squares on ordinates do not enter the argument explicitly. He produces a series of parabola properties as though they were familiar truths. These are, specifically, that $AB = BG$; that 'the line drawn from θ perpendicular to θA meets AX beyond E'; and that $GZ = BH$. All of these are true, but he did not spell out any derivation not needed for his immediate train of thought. Given the effortless and confident handling of the tangent (θA) and of what we call the *normal* (θZ, the perpendicular to the tangent at the point of tangency), we can judge that this is a sophisticated treatment by an accomplished research mathematician.

(ii) So far as the main body of the proof goes, it reads as an efficient contribution to the theory of conic sections. But the last paragraph, and the introduction **SB** 4.C1, place this theoretical knowledge in the practical context of mirror construction. So the overall enterprise has slipped beyond the grasp of the strict Platonic canons. There are several profound physical assumptions embedded in his procedure; that light rays can be treated as though they were geometrical lines, that rays from the sun are parallel, and so on. ■

The surface he described at the end, resulting from revolving the parabola around its axis, is just the paraboloid of revolution whose investigation by Archimedes you studied in Section 4.3.

When we turn to Diocles' contemporary Apollonius, we are unambivalently back in the high geometric research tradition. To attempt to work through Apollonius' *Conics* is a somewhat unnerving experience, as of wandering an endless marble labyrinth for reasons which begin to escape one. There is nothing here of the gritty excitement of Archimedes, or the relative accessibility of Euclid, still less the warm humanity and wise irony of Plato. Apollonius presents an inexorable piling-up of 387 propositions in the seven surviving Books (the eighth being lost), icily austere in forbiddingly rigorous geometrical language, the rationale for the pattern of the whole but dimly perceptible. Many subsequent writers have commented on the difficulty of grasping the principles of selection and of ordering his material, with views varying from that of Descartes who seemed to suggest that Apollonius wrote down whatever came into his head, to that of B. L. van der Waerden who observed

Descartes, *La Geometrie*, tr. D. E. Smith and M. L. Latham (Dover, 1954) pp. 16–17.

Apollonius is a virtuoso in dealing with geometric algebra, and also a virtuoso in hiding his original line of thought. This is what makes his work hard to understand; his reasoning is elegant and crystal clear, but one has to guess at what led him to reason in this way, rather than in some other way.

van der Waerden, *Science Awakening*, p. 248.

Nevertheless there are some threads we can grasp to find our way through, and may hope to emerge with nuggets of gold as testimony to this extraordinary work and its considerable influence in later centuries.

Little is known of Apollonius' life beyond what can be gleaned from the prefatory letters to the various Books in which he circulated the *Conics*. The impression given is not unlike that of other members of the research mathematical community over the previous couple of centuries, of someone who had travelled about and spent some time in various places around the Mediterranean, meeting people of like interests with whom he kept in touch. Born in Perga on the south coast of Asia Minor, north-west of Cyprus, Apollonius mentioned Alexandria, Ephesus and Pergamum as places he visited or stayed at. The Preface to Book I is of particular interest, as Apollonius there describes the plan of the whole work; please *read that now* (**SB** 4.D1).

SB 4.D1–D2

Pergamum was an important city in northern Asia Minor whose library was second only to Alexandria in repute—the word '*parchment*' comes from its name.

You may have noticed where Descartes got the idea that Apollonius put down whatever came into his mind; even Apollonius seems aware that some confusion could arise from the circulation of different versions of his work. We concentrate on topics from the first four books which he described in the Preface as 'an elementary introduction'. Book I, he said, contains the fundamental properties of the conic sections 'worked out more fully and generally than in the writings of others'. This is a crucial clue to his achievement. It seems that Apollonius did successfully for conic sections what Euclid did not quite manage (nor perhaps intend) for geometry as a whole, to rework the elements from scratch as a unified theory. It is in large measure the generality of Apollonius' treatment that makes it difficult to comprehend, yet so rewarding once one has done so. His generality was firmly announced right from the start. Look at *First Definitions* 1–3 (in **SB** 4.D3). Immediately we are confronted with a more general framework than we have met before.

Bear in mind our earlier discussion of the meaning of the word *elements*, Unit 3, Section 4; Apollonius was not claiming the books were simple or easy.

Question 12 Compare Apollonius' definition of a *cone* with that given by Euclid (*Elements* XI, Def. 18, **SB** 3.E2). How, essentially, do they differ?

Comment —————————————————————————

They differ in two major ways, and one minor one. The latter is that Euclid's cone is a figure of revolution of a right-angled triangle, whereas Apollonius has a straight line fixed at a point and tracing round a circle. But they could still amount to the same thing, were it not for the two striking features of Definition 1; the line is of indefinite length in both directions; and the point need not be vertically above the

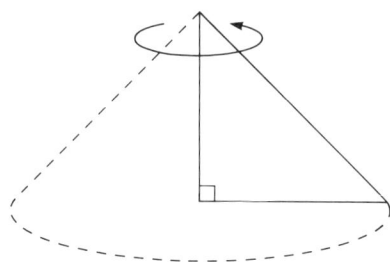

Figure 18

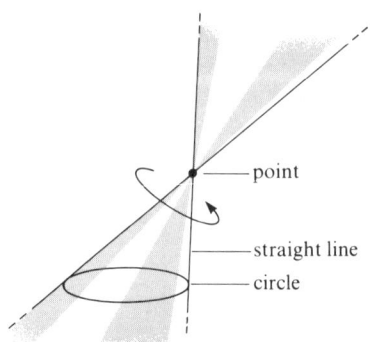

— point

— straight line

— circle

Figure 19

centre of the circle (assuming the latter to be horizontal), but can be anywhere except in the plane of the circle. So Apollonius' general starting object, the *conic surface*, is a double extended cone which can be quite skewed. His *cone* is half that, down as far as the circle; and Euclid's cone is a special case of that, what Apollonius called a *right cone* in Definition 3. ■

So clearly Apollonius is preparing us for something rather spectacular; nor are we disappointed. After a few ground-clearing propositions, much as a stallion paws the ground before charging, Apollonius unleashes three stunning propositions (11–13; **SB** 4.D4) setting the tone for the whole work. These show that taking any cone and a plane section through it, then the resultant curve has the same symptom as one of the previous known conic sections. So what were previously sections perpendicular to a side of three different cones can now be arrived at by three different sections of one cone.

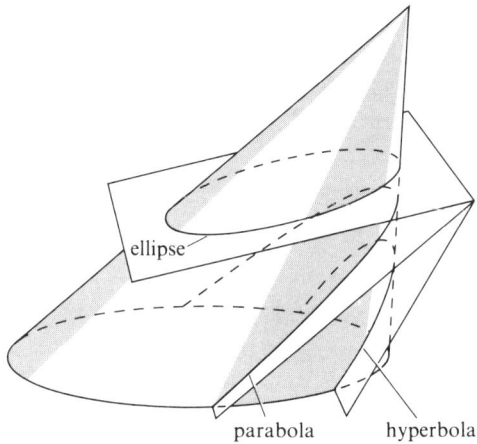

Beware a danger of confusion here with the previous terminology: in *right-angled cone*—the sort of thing we have seen whose section is a parabola—the right angle in question is that at the point of the cone; for a *right cone*, it is the angle between the axis and the base circle that is a right angle.

Figure 20

That they are indeed the same curves as before, as opposed to curves having the same symptoms, emerges during the rest of Book I. Apollonius called them *parabola*, *hyperbola* and *ellipse*. The reason for this is interesting. These terms describe not the genesis of the curves (as did the previous names 'section of an acute-angled cone' etc.), but their symptoms, reflecting the geometrically-expressed property satisfied by points on the curves. To see what this property is, remember that

(i) Apollonius is working in as general a way as possible, so any point on the curve can be considered in relation to its distance measured along some diameter (his Definition 4) and up parallel to the tangent at the end. (These are the distances x and y in Figure 21.)

(ii) Conic sections have constant *parameters* associated with them, related in the old definition to the distance of the cutting plane from the vertex of the cone.

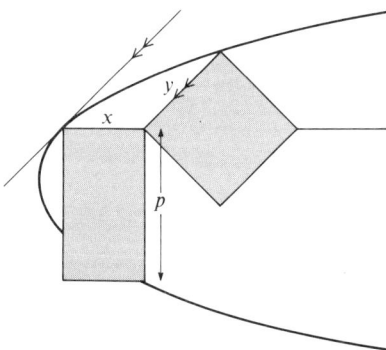

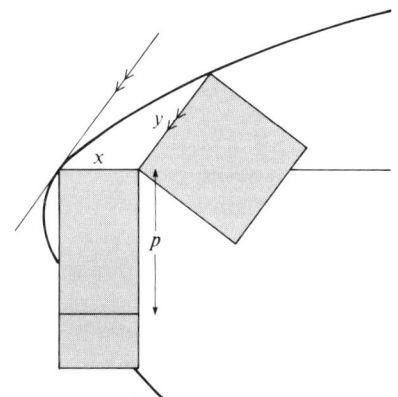

Figure 21(a) Parabola   **Figure 21(b)** Ellipse   **Figure 21(c)** Hyperbola

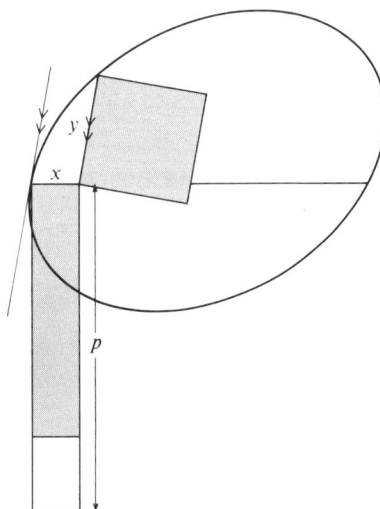

25

Apollonius showed that for each curve the two shaded areas on each diagram are equal, one being the square on the ordinate y, and the other being a rectangle one of whose sides is the abscissa x. It is how the other side of the rectangle relates to the parameter p that determines the shape and name of the curve. If the other side is p itself, then the relation is what we would express as $y^2 = px$, and the curve is a parabola. When the other side is shorter than p (by an amount depending on x), the curve is an ellipse; and when it is longer than p, the curve is a hyperbola.

We shall not go through the proofs of these results. But as a further insight into Apollonius' style, you should read through the statement of the proposition for the parabola.

Question 13   Read the statement of Proposition 11 (**SB** 4.D4) and try to match up what is said there with the description of the parabola given above. What are Apollonius' terms for what have been called y, x and p?

Comment ――――――――――――――――――――――――――――――

The ordinate y is 'any straight line which is drawn from the section of the cone to its diameter parallel to the common section of the cutting plane and of the cone's base'.

The abscissa x is 'the straight line cut off by it on the diameter beginning from the section's vertex'.

The parameter p is 'another straight line which has the ratio to the straight line between the angle of the cone and the vertex of the section that the square on the base of the axial triangle has to the rectangle contained by the remaining two sides of the triangle'.

So incorporating the shorter symbolism in place of the Apollonian terms, his proposition reads:

> If . . . (the sectioning is done such a way),
> then y will equal in square the rectangle contained by x and p.
> And let such a section be called a parabola.

The middle line of this is the relation $y^2 = px$ above. There are two morals to this. One is that Apollonius makes remarkably good sense if one can find an appropriate language to understand him. The other is that Apollonius' own understanding and geometrical intuition must have been extraordinarily powerful and deep, even at a time when verbal, rhetorical expression of mathematics was the norm. ∎

One further point on this. There were both gains and losses in Apollonius' new highly general approach. Notice for instance how complicated the parameter p is in Apollonius' formulation, compared with the simple distance along the cone previously. This price must be set against the benefits of the new unified theory.

What sort of curve properties was Apollonius interested in? He did not engage with the things investigated by Archimedes and the Eudoxan tradition (quadratures, cubatures, filling things up with polygons and so forth) but rather with another aspect of geometry: what lines go where? How do they cut? How do the lengths relate? What are the different properties and relations of tangents, diameters, normals? A host of auxiliary lines and definitions came into play, leaving readers in no doubt about the richness and diversity of properties evinced by the conic sections.

Let us look briefly at one example to see the kind of result he obtained. *Conics* IV, 9 (**SB** 4.D6(d)) explains how to draw tangents to a conic section from a point outside it. We draw through the point any two lines which cut the conic section, and divide that part of each line within the curve in a particular ratio (the 'harmonic ratio' $EL:LH = DE:HD$; and similarly for the other). Where the line through the two new points cuts the curve, there the tangents from the original point meet the curve. The proof, being by *reductio ad absurdum*, is actually rather unilluminating on what this curious but significant ratio has to do with tangents. (We are persuaded, rather, by reference to the earlier proposition *Conics* III, 37, that things could not be otherwise.) Note, though, that Apollonius proved the result for any conic section, and did not need to consider the case of each one separately. This is one of the benefits of his generality of treatment, that he was able to construct a unified body of techniques and concepts in place of a previously more diverse subject.

It is easy to remember which is which because of the cognate words in English: a *parable* is a story in which a comparison of similar things is implied; an *elliptic* remark is one falling short of what is meant; and *hyperbole* is rhetorical exaggeration, exceeding what is meant.

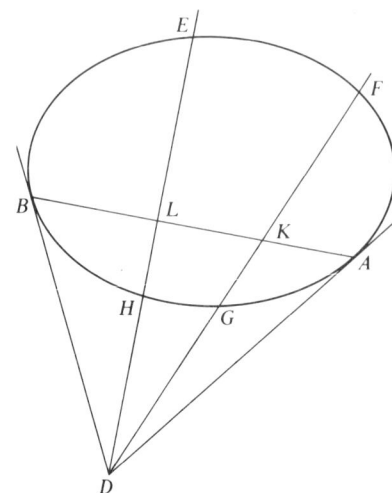

Figure 22

Finally, what is the point of all this? Had the massive edifice of Apollonius' *Conics* any particular use? This is a question that could be asked about other things too, but is raised here because Apollonius himself did so.

Question 14   Please read the Prefaces to *Conics* IV and V. (**SB** 4.D2(b) and (c)). What answers did Apollonius give to this question?

Comment ———————————————————————————————

Towards the end of each preface he gives two kinds of answer. There is the use that other mathematicians will have, to further their own studies; and there is an interesting 'art for art's sake' argument, that his theorems are intrinsically worthwhile. This aesthetic response is one echoed by many mathematicians down the ages, entirely convincing to those who see it thus, incomprehensible to those who do not. ■

With Apollonius' *Conics* your substantive exploration of Greek geometry draws to a close. The Greek tradition had many more centuries yet to roll, as you saw towards the end of *Unit 2*, but never recaptured the heights of mathematical creativity of the century which saw both Archimedes and Apollonius. Both revered figures in the centuries after their deaths, they were no less so as their works became available for sixteenth- and seventeenth-century mathematicians to study. 'He who understands Archimedes and Apollonius' said G. W. Leibniz in the late seventeenth century, 'will admire less the achievements of the foremost men of later times'.

Cited in Carl B. Boyer, *A History of Mathematics* (Wiley, 1968) p. 172.

Figure 23   Frontispiece of the 1710 edition of Apollonius' *Conics*, edited by Edmond Halley. The picture illustrates the quotation, which is from Vitruvius' *De Architectura*, Book VI: 'Aristippus the Socratic philosopher was shipwrecked on the coast at Rhodes, and, observing geometrical diagrams drawn on the sand, is said to have cried out to his companions: "Be of good cheer! For I see the traces of man".'

FURTHER READING

Allman, George J., *Greek Geometry from Thales to Euclid* (reprinted by Arno Press, 1976).
  Although now a century old (originally published in 1889), this remains a most perceptive work carefully distinguishing source material from its interpretation.

Fowler, David H., *The Mathematics of Plato's Academy* (Oxford University Press, 1987).
  An illuminating attempt to reconstruct the development of fourth-century ideas of ratio and proportion, attentive to sources and to papyrological evidence.

Heath, Sir Thomas L., *A History of Greek Mathematics* (Dover, 1981) (original edition 1921).
  The standard two-volume general history. Very thorough and informative, though some judgements need revision in the light of subsequent scholarship.

Knorr, Wilbur R., *The Evolution of the Euclidean Elements* (Reidel, 1975).
  A substantive study of the development and significance of incommensurable magnitudes, which we have only touched on in this course.

Knorr, Wilbur R., *The Ancient Tradition of Geometric Problems* (Birkhäuser, 1986).
  Takes discussion of the problem-solving aspect of the geometric research tradition—notably the three classical problems—further and more fully than we do here.

Lloyd, G. E. R., *Magic, Reason and Experience* (Cambridge University Press, 1979).
  The development not only of mathematics but also of other Greek science in the context of social and intellectual history.

van der Waerden, B. L., *Science Awakening* (Oxford University Press, 1961).
  Livelier than Heath, and only slightly less thorough.